YOUR KNOWLEDGE HAS VALUE

- We will publish your bachelor's and
  master's thesis, essays and papers

- Your own eBook and book -
  sold worldwide in all relevant shops

- Earn money with each sale

Upload your text at www.GRIN.com
and publish for free

Seema Kumari

Radiation induced expression of signaling molecules in mouse splenocytes

GRIN Verlag

Bibliografische Information der Deutschen Nationalbibliothek:

Die Deutsche Bibliothek verzeichnet diese Publikation in der Deutschen National-
bibliografie; detaillierte bibliografische Daten sind im Internet über http://dnb.d-
nb.de/ abrufbar.

Imprint:

Copyright © 2004 GRIN Verlag GmbH
Druck und Bindung: Books on Demand GmbH, Norderstedt Germany
ISBN: 978-3-656-38466-3

RADIATION INDUCED EXPRESSION OF SIGNALLING MOLECULES

By SEEMA KUMARI

Radiation induced expression of signalling molecules

Authors:

Seema kumari (Department of Biochemistry, GITAM University)

Malla Ramarao and Murali Mohan.Ch (Department of Biochemistry, GITAM University)
Malini Krishna and Anirban Mitra (Department of Radiation Biology, BARC Mumbai)

Corresponding author
Seema kumari

Preface

The present research work was undertaken as M.Sc project at Radiation Biology Division, Bhabha Atomic Research Centre, Mumbai under the guidance of Dr. Malini Krishna. This work emphasises on expression levels of signalling molecules regulating cell death and survival in presence of radiation stress in radiosensitive organ spleen. This work has not been submitted to any other Universities for the award of any degree or diploma and as research work.

Seema Kumari

*This work is dedicated to my mother **Raj Rani Devi** and my father **Yogendra Prasad Singh**.*

ACKNOWLEDGEMENT

I express my profound gratitude to my teachers for their support to complete my project. I would also like to express my thanks to T.S.R & T.B.K P.G college management. I would like to thanks Dr. K.P Mishra, BARC for his support during my research work. Would like thank the authors, for their references I have used to prepare this write up.

I would like to thank god for giving me the courage my parents, my brother, sister and husband for their support and giving me the strength, courage and determination to triumph through the difficult times.

(Seema Kumari)

INDEX

Introduction

The hazards of exposure to ionizing radiation were recognised shortly after Roentgen's discovery of x-rays in 1895. Acute skin cancer, leukaemia and other biological damage were observed in the individuals working with x-ray generator. In the year, 1898 Becquerel performed the first recorded experiment in radiobiology, from this earlier study of radiobiology began. Since that time, a tremendous amount of research has been done attempting to interpret the reactions which take place from the moment that radiation enters a living cell until some permanent damage is produced. From beginning to end, these initial reactions are probably completed in a millionth of a second, making them very difficult to study. For this reason, it is still not known which of the many chemical or biochemical reactions brought about by ionizing radiation are responsible for initiating biological damage.

Ionizing radiation is energy transmitted by X-rays, gamma rays, beta particles (high-speed electrons), alpha particles (the nucleus of the helium atom), neutrons, protons, and other heavy ions such as the nuclei of argon, nitrogen, carbon, and other elements. X-rays and gamma rays are electromagnetic waves like light, but their energy is much higher than that of light (their wavelengths are much shorter). Ultraviolet (UV) light is a radiation of intermediate energy that can damage cells like sunburns, but UV light differs from the forms of electromagnetic radiation mentioned above in that it does not cause ionization (loss of an electron) in atoms or molecules, but rather excitation (change in energy level of an electron). The other forms of radiation particles are either negatively charged (electrons), positively charged (protons, alpha rays, and other heavy ions), or electrically neutral (neutrons).

Ionization

As an example of ionization, beta rays are fast electrons that lose energy as they pass through cells and interact with molecules. The transferred energy is high enough to disrupt chemical bonds, which results in radical formation (or ionization). When an electron passes through a cell, it releases its energy along its path (called a track) by interacting with the electrons of nearby molecules. The released energy is absorbed by atoms near the track, resulting in either excitation (a shift in the orbit of an electron to a higher energy level) or ionization (release of an electron from the atom), which makes the atoms very unstable. Such unstable atoms are called radicals and are chemically very reactive. Some radicals are so reactive that they exist only for as short a time as a microsecond. X-ray and gamma rays differ from beta particles in that they release high-speed electrons from atoms first. Positively charged particles transfer energy to molecules in cells by essentially the same mechanisms. Neutrons are somewhat different since they are electrically uncharged, and their main effect is to impact the nuclei of hydrogen atoms, namely protons. Since the masses of a neutron and a proton are similar, the impact results in an elastic scattering process like in billiards. The ejected protons behave as charged particles.

Effect of ionizations on cells

Radiation-induced ionizations may act directly on the cellular component molecules or indirectly on water molecules, causing water-derived radicals. Radicals react with nearby molecules in a very short time, resulting in breakage of chemical bonds or oxidation (addition of oxygen atoms) of the affected molecules. The major effect in cells is DNA breaks. Since DNA consists of a pair of complementary double strands, breaks of either a single strand or both strands can occur. However, the latter is believed to be much more important biologically. Most single-strand breaks can be repaired normally, due to double-stranded nature of the DNA molecule (the two strands complement each other, so that an intact strand can serve as a template for repair of its damaged, opposite strand). In the case of double-strand breaks, however, repair is more difficult and erroneous re-joining of broken ends may occur. These so-called misrepairs result in induction of mutations, chromosome aberrations, or cell death.

Characteristics of DNA damage by radiation exposure

Deletion of DNA segments is the predominant form of radiation damage in cells that survive irradiation. It may be caused by (1) misrepair of two separate double-strand breaks in a DNA molecule with joining of the two outer ends and loss of the fragment between the breaks or (2) the process of cleaning (enzyme digestion of nucleotides, the component molecules of DNA) of the broken ends before re-joining to repair one double-strand break.

Biological effects differ by type of radiation

Radiations differ not only by their constituents (electrons, protons, neutrons) but also by their energy. Radiations that cause dense ionization along their track (such as neutrons) are called high-linear-energy-transfer (high-LET) radiation. Low-LET radiations produce ionizations only sparsely along their track hence, almost homogeneously within a cell. Radiation dose is the amount of energy per unit of biological material (i.e. number of ionizations per cell). Thus, high-LET radiations are more destructive to biological material than low-LET radiations-- such as X rays and gamma rays, at the same dose, the low-LET radiations induce the same number of radicals more sparsely within a cell, whereas the high-LET radiations such as neutrons and alpha particles transfer most of their energy to a small region of the cell. The localized DNA damage caused by dense ionizations from high-LET radiations is more difficult to repair than the diffuse DNA damage caused by the sparse ionizations from low-LET radiations (Radiation effects research foundation).

Direct effect

If radiation interacts with the atoms of the DNA molecule, or some other cellular component critical to the survival of the cell, it is referred to as a direct effect. Such an interaction may affect the ability of the cell to reproduce and thus, survive. If enough atoms are affected such that the chromosomes do not replicate properly, or if there is significant alteration in the information carried by the DNA molecule, then the cell may be destroyed by "direct" interference with its life-sustaining system.

Indirect effect

If a cell is exposed to radiation, the probability of the radiation interacting with the DNA molecule is very small since these critical components make up such a small part of the cell. However, each cell, just as is the case for the human body, is mostly water. Therefore, there is a much higher probability of radiation interacting with the water that makes up most of the cell's volume. When radiation interacts with water, it may break the bonds that hold the water molecule together, producing fragments such as hydrogen (H) and hydroxyls (OH). These fragments may recombine or may interact with other fragments or ions to form compounds, such as water, which would not harm the cell. However, they could combine to form toxic substances, such as hydrogen peroxide (H_2O_2), which can contribute to the destruction of the cell.

Cellular sensitivity to radiation

Not all living cells are equally sensitive to radiation. Those cells which are actively reproducing are more sensitive than those which are not. This is because dividing cells require correct DNA information in order for the cell's offspring to survive. A direct interaction of radiation with an active cell could result in the death or mutation of the cell, whereas a direct interaction with the DNA of a dormant cell would have less of an effect. As a result, living cells can be classified according to their rate of reproduction, which also indicates their relative sensitivity to radiation. This means that different cell systems have different sensitivities. Lymphocytes (white blood cells) and cells which produce blood are constantly regenerating, and are, therefore, the most sensitive. Reproductive and gastrointestinal cells are not regenerating as quickly and are less sensitive. The nerve and muscle cells are the slowest to regenerate and are the least sensitive cells. Cells, like the human body, have a tremendous ability to repair damage. As a result, not all radiation effects are irreversible. In many instances, the cells are able to completely repair any damage and function normally. If the damage is severe enough, the affected cell dies. In some instances, the cell is damaged but is still able to reproduce. The daughter cells, however, may be lacking in some critical life-sustaining component, and they die. The other possible result of radiation exposure is that the cell is affected in such a way that it does not die but is simply mutated. The mutated cell reproduces and thus perpetuates the mutation. This could be the beginning of a malignant tumor.

Organ sensitivity to radiation

The sensitivity of the various organs of the human body correlate with the relative sensitivity of the cells from which they are composed. For example, since the blood forming cells were one of the most sensitive cells due to their rapid regeneration rate, the blood forming organs are one of the most sensitive organs to radiation. Muscle and nerve cells were relatively insensitive to radiation, and therefore, so are the muscles and the brain.

The rate of reproduction of the cells forming an organ system is not the only criterion determining overall sensitivity. The relative importance of the organ system to the well being of the body is also important. One example of a very sensitive cell system is a malignant tumor. The outer layer of cells reproduces rapidly, and also has a good supply of blood and oxygen. Cells are most sensitive when they are reproducing, and the presence of oxygen increases sensitivity to radiation. Anoxic cells (cells with insufficient oxygen) tend to be inactive, such as the cells located in the interior of a tumor. As the tumor is exposed to radiation, the outer layer of rapidly dividing cells is destroyed, causing it to "shrink" in size. If the tumor is given a massive dose to destroy it completely, the patient might die as well. Instead, the tumor is given a small dose each day, which gives the healthy tissue a chance to recover from any damage while gradually shrinking the highly sensitive tumor. Another cell system that is composed of rapidly dividing cells with a good blood supply and lots of oxygen is the developing embryo. Therefore, the sensitivity of the developing embryo to radiation exposure is similar to that of the tumor, however, the consequences are dramatically different. Whole body sensitivity depends upon the most sensitive organs which, in turn, depend upon the most sensitive cells. As noted previously, the most sensitive organs are the blood forming organs and the gastrointestinal system. The biological effects on the whole body from exposure to radiation will depend upon several factors. Some of these are listed above. For example, a person, already susceptible to infection, who receives a large dose of radiation, may be affected by the radiation more than a healthy person.

Effect of radiation dose

Biological effects of radiation are typically divided into two categories. The first category consists of exposure to high doses of radiation over short periods of time producing acute or short term effects. The second category represents exposure to low doses of radiation over an extended period of time producing chronic or long term effects. High doses tend to kill cells and damage tissue and organs, this in turn may cause a rapid whole body response often called the Acute Radiation Syndrome (ARS). While low doses tend to damage or change over long periods of time don't cause an immediate problem to any body organ (Biological Effects of Radiation).

Biological Response to High Doses of Radiation (Biological Effects of Radiation)

< 5 rad — - No immediate observable effects

~ 5 rad to 50 rad — - Slight blood changes may be detected by medical evaluations.

~ 50 rad to 150 rad — - Slight blood changes will be noted and symptoms of nausea,fatigue, vomiting, etc. likely.

~ 150 rad to1,100 rad - Death due to destruction of blood forming organs.

~ 1,100 rad to 2,000 rad — -Death due to destruction of nervous system and GI

Effect of low dose

There are three general categories of effects resulting from exposure to low doses of radiation. These are:

Genetic   - The effect is suffered by the offspring of the individual exposed.

Somatic   - The effect is primarily suffered by the individual exposed. Since cancer is the primary result, it is sometimes called the carcinogenic effect.

In-Utero   - Some mistakenly consider this to be a genetic consequence of radiation exposure, because the effect, suffered by a developing embryo/ fetus, is seen after birth. However, this is actually a special case of the somatic effect, since the embryo/fetus is the one exposed to the radiation.

Signalling molecules

a) PKC: Twelve serine/threonine protein kinases constitute the Protein Kinase C (PKC) family which transduces a myriad of signals which regulates multiple physiological functions (Hug and Sarre, 1993). Three distinct subfamilies of PKC isoforms are identified according to their dependency on three combinations of activators:

Conventional PKCs (α, ßI, ßII, γ) require phosphatidylserine (PS), diacylglycerol (DAG), and Ca^{2+}. Novel PKCs (δ, ε, η, θ) need PS and DAG but not Ca^{2+}. A typical PKCs (z, i/λ, μ) are insensitive to both DAG and Ca^{2+} (Hug and Sarre, 1993; Newton, 1997). The primary structure of PKC reveals the presence of four domains conserved across PKC isoforms (C1-C4) and five variable domains that are divergent (V1-V5). Two functional domains have been described in PKC: an amino terminal regulatory domain and a carboxyl terminal catalytic domain (Hug and Sarre, 1993). The regulatory domain (V1-V3) contains the so-called pseudo substrate site which is thought to interact with the catalytic domain to retain PKC in an inactive conformation. The regulatory domain also contains sites for the interaction of PKC with PS, DAG/phorbol ester, and Ca^{2+}. The Ca^{2+} dependency is mediated by the C2 region (which is indeed absent in novel PKCs), while phorbol-ester binding requires the presence of two cysteine-rich zinc-finger regions within the C1 domain. Atypical PKCs lack one of the two cysteine-rich zinc-finger regions and therefore do not bind (and cannot be activated by) phorbol esters (Hug and Sarre, 1993). The catalytic domain (V3-V5) contains ATP binding site and is thought to interact with the substrates (Hug and Sarre, 1993). Fatty acids, D3- phosphorylated inositol lipids, phosphatidic acid (PA), and ceramide are additional molecules that are considered to be capable of activating PKC (Newton, 1997; Liu and Heckman, 1998).

When activated by biomechanical stress or neuro hormonal mediators, G-protein coupled receptors separate heterotrimeric G-proteins to G-protein alpha-q/11 subunits and heterodimeric G-protein beta/gamma subunits. G-proteins bind and activate Phospholipase C beta (PLC-beta), recruit PLC-beta to the membrane where it hydrolyses Phosphatidylinositol 4,5 bisphosphate (PtdIns (4,5)P2) and releases Inositol 1,4,5-triphosphate (IP3) and DAG. IP3 binds to receptors (IP3R) in the endoplasmic reticulum, releasing calcium Ca $^{2+}$. The increase in cytosolic Ca $^{2+}$ activates the protein phosphatase calcineurin. Calcineurin dephosphorylates several residues in the amino-terminal region of the transcription factor NF-AT, allowing it to translocate to the nucleus and activate transcription of hypertrophic response genes.

PKC-alpha, PKC-delta, PKC-epsilon, PKC-zeta and PKC-mu phosphorylate and activate PKC-potentiated inhibitor protein of 17kDa (CPI-17). CPI-17 specifically inhibits myosin light chain phosphatase (MLCP), leading to MELC phosphorylation by MLCK. MLCK in turn is activated by calmodulin (Zemlickova et al., 2004). One of the PKC-regulated pathways leads to the inhibition of a subset of Histone deacetylases (HDAC7) that specifically regulate cellular hypertrophy. In this pathway, PKC-delta activates another protein kinase, PKC-mu, that in its turn phosphorylates the HDAC7 leading to its export from the nucleus and consequent inactivation (Li et al., 2004).

PKC-mu activates the transcription factor Nuclear factor kappaB (NF-kB). PKC-mu phosphorylates the IKK beta, leading to I-kB degradation and subsequent NF-kB translocation into the nucleus (Storz et al.,2003). Activation of PKC-mu in response to oxidative stress requires its sequential phosphorylation by two kinases, tyrosine kinase cABL and PKC-delta (Storz et al., 2004). PKC-mu activation leads to the transcriptional activation of NUR77 via Myocyte enhancer factor 2 (MEF2)-binding sites in its promoter (Parra et al., 2005). v-Src sarcoma viral oncogene homolog (c-Src) phosphorylates and activates PKC-iota (Wooten et al.,2001). Atypical PKC-zeta is activated by Ceramide. This results in activation of NF-kB and continued survival of the cell (Wang et al., 1999).

The two members of the atypical protein kinase C (aPKC) subfamily of isozymes (PKC-zeta and PKC-iota) are involved in control of the NF-kB activity through IKKbeta activation. A PKC-binding protein Sequestosome 1(p62) selectively interacts with receptor-interacting protein RIPK1 as an adaptor. Sequestosome 1(p62) bridges atypical PKCs and RIPK1. The latter activates IKK gamma, and atypical PKCs phosphorylate and activate IKKbeta. Thereby, the interactions of Sequestosome 1(p62) with RIPK1 and the atypical PKCs lead to the activation of NF-kB signalling pathway (Sanz et al., 1999). The PKC-theta isoform also induces NF-kB activation. PKC-theta directly targets IKK beta for phosphorylation and activation, possibly via homodimeric IKKbeta complexes (Altman et al., 2003). PKC-alpha, PKC-beta, PKC-gamma, PKC-epsilon, and PKC-eta phosphorylate and activate v-Raf-1 murine leukemia viral oncogene homolog 1 (c-Raf-1) leading to the stimulation of the Mitogen-activated protein kinase kinase 1 and 2 (MEK1 and MEK2)/ Mitogen-activated protein kinases 1 and 3 (ERK1/2) cascade and activation of the transcription factor Elk-1 (Hamilton et al., 2001).

Several PKC isotypes (PKC-alpha, PKC-beta, PKC-gamma, PKC-delta, and PKC-eta) phosphorylate Glycogen synthase kinase 3 beta (GSK3-beta) and inactivate it (Fang et al., 2002). GSK3-beta phosphorylates conserved serines of NF-AT. This phosphorylation promotes the nuclear exit of NF-AT, thereby opposing Ca $^{2+}$ - calcineurin signalling (Dorn et al., 2005).

b) Protein kinase C-delta (PKCδ): A member of the lipid-regulated serine/threonine PKC family has been implicated in a wide range of important cellular processes. In the past decade, the critical role of PKCδ in regulation of both intrinsic and extrinsic apoptosis pathways has been widely explored. In most cases, over-expression or activation of PKCδ results in the induction of apoptosis. The phosphorylations and multiple cell organelle translocations of PKCδ initiate apoptosis by targeting multiple downstream effectors. During apoptosis, PKCδ is proteolytically cleaved by caspase-3 to generate a constitutively activated catalytic fragment, which amplifies apoptosis cascades in nucleus and mitochondria. However, PKCδ also exerts its anti-apoptotic and pro-survival roles in some cases. Therefore, the complicated role of PKCd in apoptosis appears to be stimulus and cell type dependent (Meng et al., 2012).

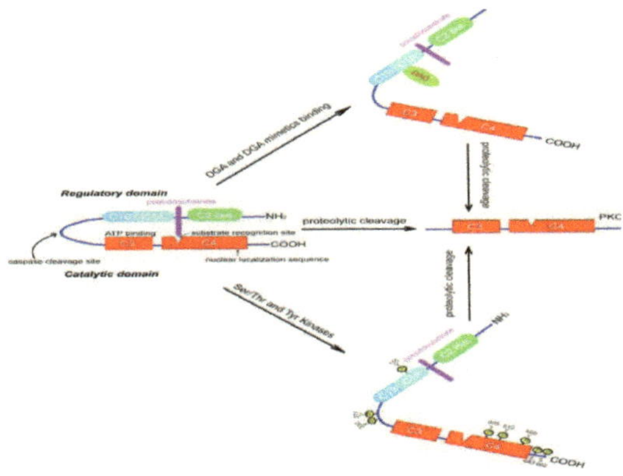

Figure. 1 Diagramatic activation mechanisms of PKC δ. (*Ref: Meng Zhao • Li Xia • Guo-Qiang Chen Protein Kinase Cd in Apoptosis: A Brief Overview Arch. Immunol. Ther. Exp. (2012) 60:361–372*).

PKCδ binds to DAG and it is phosphorylated by Ser/Thr and Tyr kinases, which changes the conformation and exposes the substrate binding site of PKC δ to get activated. PKCδ is proteolytically cleaved by caspase-3, which releases the catalytic fragment, a constitutively active form.

c) MAPKs (Mitogen-Activated Protein Kinases)

These are Serine-threonine protein Kinases that are activated in response to a variety of extracellular stimuli and mediate signal transduction from the cell surface to the nucleus. MAPKs are expressed in multiple cell types including cardiomyocytes, vascular endothelial cells, and vascular smooth muscle cells. Three major MAPKs include ERKs (Extracellular signal-Regulated Kinases), JNKs (c-Jun NH (2)-terminal protein Kinases), and p38 Kinases. Members of the JNK/SAPK (Stress-Activated Protein Kinase) family of MAPKs are strongly stimulated by numerous environmental stresses, but also more modestly stimulated by mitogens, inflammatory cytokines, oncogenes, and inducers of cell differentiation and morphogenesis. Ten mammalian JNK isoforms have been identified and are encoded by three distinct genes, JNK1, JNK2, and JNK3, the transcripts of which are alternatively spliced to yield four JNK1 isoforms, four JNK2 isoforms, and two JNK3 isoforms. JNK1 and JNK2 are the products of alternative splicing of a single gene and are expressed in many tissues, but JNK3 is specifically expressed in brain. Members of the JNK family play crucial roles in regulating responses to various stresses, and in neural development, inflammation, and apoptosis. JNK activation is much more complex than that of ERK1/ERK2 owing to inputs by a greater number of MAPKKKs (Mitogen-Activated Protein Kinase Kinase Kinases) at least 13, including MEKK1 (MAP/ERK Kinase-Kinase-1)-MEKK4 (MAP/ERK Kinase-Kinase-4), ASK (Apoptosis Signal-regulating Kinase) and MLKs (Mixed-Lineage Kinases), which are activated by upstream Rho-family GTPases. These activate JNK MAPKKs MEK4 (MAPK/ERK Kinase-4) and MEK7 (MAPK/ERK Kinase-7), which further activate JNKs. The JNK MAPK modules are regulated by a number of different scaffold proteins, including JIP1 (JNK Interacting Protein-1), JIP2 (JNK Interacting Protein-2), JIP3 (JNK Interacting Protein-3), JIP4 (JNK Interacting Protein-4), Beta-Arrestin-2, Filamin and CrkII. The scaffold proteins presumably target the MAPK modules to different sites in the cell and play roles in kinase activation and/or substrate selection (Himes et al., 2006 & Moulin et al., 2004).

Stress or genotoxic agents are the most powerful inducers of JNK. Different forms of stress have been shown to mediate JNK activation via various cellular pathways. JNK activation in response to UV irradiation is mediated by upstream signalling components, including Rac (Ras-Related C3 Botulinum Toxin Substrate), CDC42 (Cell Division Cycle-42), PAK (p21/CDC42/Rac1-Activated Kinase), ASK1 (Apoptosis Signal-regulating Kinase-1), MLK, MEKK1, SEK1 (SAPK/ERK Kinase-1)/MKK4, MKK7 and p21Ras, in concert with nuclear DNA lesions. Besides stress, JNKs can also be activated via GPCRs (G-Protein Coupled Receptors), RTKs (Receptor Tyrosine Kinases) and cytokine receptors. How GPCRs activate the JNKs is still an unanswered question. Free Beta-Gamma dimers and GN-Alpha12 and GN-Alpha13 proteins are able to activate JNK in a Rac1-CDC42 or p115RhoGEF and RhoA-dependent manner. However, the nature of the GEFs (Guanine nucleotide Exchange Factors) that connect Beta-Gamma and GN-Alpha12/ GN-Alpha13 to Rac1 and CDC2 is still unclear. Interestingly, GN-Alpha12 can also activate JNK by activating the MEKK (MEK kinase). The activation of JNK by cytokine receptors appears to be mediated by the TRAF (TNF Receptor-Associated Factor) group of adaptor proteins. Activation of the TNF receptor leads to recruitment of TRAF2 (TNF Receptor-Associated Factor-2), which is required for JNK activation. These adaptor proteins (TRADD (Tumor Necrosis Factor Receptor-1-Associated Death Domain Protein), RIP (Receptor-Interacting Protein) and Daxx) have been reported to bind MEKK1 and ASK1. TRAF2 activates MAPK4Ks like GCK (Germinal Center Kinase), GCKR (GCK-Related Kinase), GLK (GCK-Like Kinase) and HGK (HPK/GCK-like Kinase), which further activates JNKs via MEKK1 and MKK4/7 respectively. ASK1 also interacts with TRAF2 and activates JNK via MKK4/7 (Zhou et al., 2006 and Yang et al., 2006).

Growth factors also activate JNKs. Although the signalling cascade from growth factor receptors to ERKs is relatively well understood, the pathway leading to JNK activation is more obscure. Activation of JNK by EGF (Epidermal Growth Factor) or NGF (Nerve Growth Factor) is dependent on H-Ras activation. Growth factors and growth factor receptors stimulate Ras by recruiting SOS (Son of Sevenless), GRB2 (Growth Factor Receptor-Bound Protein-2) and SHC to the membrane. PI3K (Phosphatidylinositde-3-Kinase) also activate Ras. Ras activates two protein kinases, Raf1 and MEKK (MEK (MAPK, or ERK, kinase)

Kinase). Raf1 contributes directly to ERK activation but not to JNK activation, whereas MEKK participated in JNK activation but caused ERK activation only after overexpression. Recently, Raf1 is found to interact with the proapoptotic, stress-activated protein kinase ASK1 in vitro and in vivo. This interaction allows Raf1 to act independently of the MEK–ERK pathway to activate JNK pathway (Kraus et al., 2003 & Matsukawa et al., 2004). The Rho family GTPases, CDC42 (Cell Division Cycle-42) and Rac also initiate a cascade leading to JNK/SAPK, presumably by binding and activating the protein kinase PAK (p21-Activated Kinases), a kinase that phosphorylates and promotes activation of MEKK1. Rac/CDC42 are also involved in JNK activation via a pathway consisting of a sequential cascade MLKs and MKK4/7 (MAP Kinase Kinase-4/7. MLK2 (Mixed-Lineage Kinase-2) and MLK3 (Mixed-Lineage Kinase-3) interact with the activated (GTP-bound) forms of Rac and CDC42, with a slight preference for Rac. Besides MLKs, MEKK1/4 and Posh (Plenty of SH3) are also activated by Rac/CDC42 to activate MKK4/7 and thus JNKs. Adaptor proteins such as Crk (v-Crk Avian Sarcoma Virus Ct10 Oncogene Homolog) and CrkL (v-Crk Avian Sarcoma Virus Ct10 Oncogene Homolog-Like) also leads to activation of JNKs in response to RTK. HPK1 (Hematopoietic Progenitor Kinase-1) associates with Crk and CrkL through binding to the SH3 (Src-Homology Domain-3) of these proteins. Furthermore, association of HPK1 with these proteins increases HPK1's kinase activity. HPK1 then act as upstream of MEKK1 and TAK1 (Transforming Growth Factor-Beta-activated Kinase-1) in the JNK kinase cascade. JNKs are negatively regulated by MKP (MAP Kinase Phosphatase) (Yamauchi et al., 2002 & Zhou et al., 2006).

The activated JNK/SAPKs translocate to the nucleus where they phosphorylate transcription factors such as c-Jun, Elk1, DPC4 (Deleted In Pancreatic Carcinoma 4)/ SMAD4 (Sma and MAD (Mothers Against Decapentaplegic) Related Protein-4), p53, ATF2 (Activating Transcription Factor-2), NFAT4 (Nuclear Factor of Activated T-Cell-4) and NFAT1 (Nuclear Factor of Activated T-Cell-1). JNK1 directly phosphorylates Bcl2 (B-Cell CLL/Lymphoma-2) in vitro, co-localizes and collaborates with Bcl2 to mediate prolonged cell survival. JNK cascade also activates TCF (Ternary Complex Factor) protein. JNK also phosphorylate HSF1 (Heat Shock Factor-1) and JNK-mediated phosphorylation of HSF1 selectively stabilize the HSF1 protein and confers protection to cells under conditions of severe stress. DCX is also a substrate of JNK and interacts with both JNK and JIP. MAPs (Microtubule-Associated Proteins), both MAP1B and MAP2B are also found to be the substrates of JNK. Ser-727 phosphorylation of STAT3 (Signal Transducer and Activator of Transcription-3) can also be induced by JNK. JNK also regulates Insulin signalling by negatively regulating IRS1 (Insulin Receptor Substrate-1). JNK is generally thought to be involved in inflammation, proliferation and Apoptosis. Accordingly, its substrates are transcription factors and Anti-apoptotic proteins. However, JNK also phosphorylates Serine 178 on Paxillin and regulate cell migration. Despite extensive progress in the understanding of the JNK MAP kinase pathway, the mechanisms by which the pathway contributes to the many cellular programs where JNKs are activated are poorly defined. The JIP1 proteins have been proposed to act as molecular scaffolds that organize the JNK signal transduction pathway in response to specific stimuli. The JNK stress pathway is thought to be important in many pathological conditions including the progression of some neurodegenerative diseases such as Huntington's and also in cancer. This pathway therefore offers potential targets for therapeutic intervention. The identification of critical components of this signalling pathway, such as JIP1, offers new routes to understand how this pathway is regulated and potential ways of manipulating it to combat disease (Baan et al., 2006 Sprowles 2005& Heasley et al., 2006).

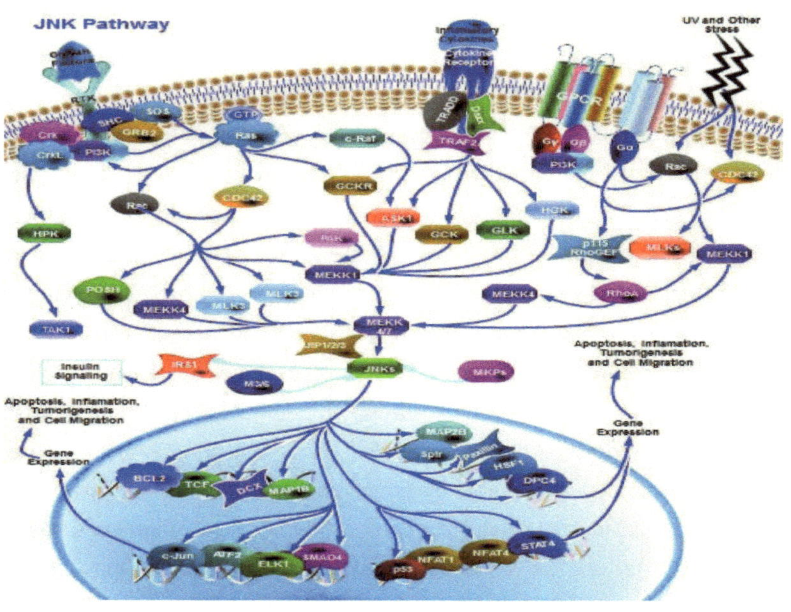

Figure 2. Diagrammatic representation of JNK pathway (*Reference: SA Bioscience*)

d) P44/42

Mitogen-activated protein kinases (MAPKs) are a widely conserved family of serine/threonine protein kinases involved in many cellular programs, such as cell proliferation, differentiation, motility, and death. The p44/42 MAPK (Erk1/2) signalling pathway can be activated in response to a diverse range of extracellular stimuli including mitogens, growth factors, and cytokines (Meloche et al., 2007), and research investigators consider it an important target in the diagnosis and treatment of cancer (Meloche et al., 2007). Upon stimulation, a sequential three-part protein kinase cascade is initiated, consisting of a MAP kinase kinase kinase (MAPKKK or MAP3K), a MAP kinase kinase (MAPKK or MAP2K), and a MAP kinase (MAPK). Multiple p44/42 MAP3Ks have been identified, including members of the Raf family, as well as Mos and Tpl2/COT. MEK1 and MEK2 are the primary MAPKKs in this pathway (Murphy et al.,

2006). MEK1 and MEK2 activate p44 and p42 through phosphorylation of activation loop residues Thr202/Tyr204 and Thr185/Tyr187, respectively. Several downstream targets of p44/42 have been identified, including p90RSK (Dalby et al., 1998) and the transcription factor Elk-1 (Kortenjann et al., 1994). p44/42 are negatively regulated by a family of dual-specificity (Thr/Tyr) MAPK phosphatases, known as DUSPs or MKPs (Owens et al., 2007), along with MEK inhibitors, such as U0126 and PD98059.

e) p38

JNK and p38 are stress-activated MAPkinases that are preferentially activated by cell stress-inducing signals, including oxidative stress, environmental stress, and toxic chemical insults. Sustained activation of JNK or p38 is implicated in the induction of many forms of neuronal apoptosis in response to a variety of cellular injuries. Apoptosis induced by cellular stress is often mediated through the mitochondria initiated cell death pathway. The Bcl-2 family proteins regulate this process by modulating the membrane potential and function of mitochondria. There has been intense interest in understanding how both pro- and anti-apoptotic kinase-signalling pathways regulate the function of Bcl-2-related proteins. For example, the activities of many Bcl-2 family proteins, including BAD, Bcl-2, and Bcl-xL, are regulated by protein phosphorylation. Bim is a BH3 domain-only pro-apoptotic protein and a member of the Bcl-2 family with three major forms generated by alternative splicing: BimEL, BimL, and BimS. BimEL is the most abundant isoform in neurons. Recent studies suggest that JNK induces apoptosis by directly phosphorylating BAD, BimEL, and BimL. In addition, JNK also phosphorylates and inactivates the anti-apoptotic Bcl-2 and Bcl-xz. In contrast to extensive studies concerning the regulation of Bcl-2 family members by JNK, there is no evidence that p38 regulates apoptosis through direct phosphorylation of Bcl-2 family proteins. (Beibei et al., 2006).

Cytoprotective pathways

The Akt and mitogen-activated protein kinase (MAPK) pathways have been implicated in tumor cell survival and contribute to radiation resistance. However, the molecular basis for link between MAPK and Akt in cell survival response to radiation is unclear. Ionizing radiation triggered Thr(308) and Ser(473) phosphorylation of Akt. Exposure of cells to radiation also induced p38 MAPK and c-Jun NH(2)-terminal kinase activations. Inhibition of c-Jun NH(2)-terminal kinase suppressed radiation-induced cell death, whereas inhibition of p38 MAPK effectively increased sensitivity to radiation. Interestingly, inhibition of p38 MAPK completely attenuated radiation-induced Ser(473) phosphorylation of Akt but did not affect Thr(308) phosphorylation. Conversely, overexpression of p38 MAPK enhanced Ser(473) phosphorylation of Akt in response to radiation. In addition, inhibition of p38 MAPK failed to alter phosphoinositide 3-kinase and phosphoinositide-dependent protein kinase activities. Ectopic expression of RacN17, dominant-negative form of Rac1, inhibited p38 MAPK activation and Ser(473) phosphorylation of Akt. Following exposure to radiation, c-Src was selectively activated among Src family tyrosine kinases. Inhibition of c-Src attenuated Rac1 and p38 MAPK activations and Ser(473) phosphorylation of Akt. The c-Src-Rac1-p38 MAPK pathway is required for activation of Akt in response to radiation and plays a cytoprotective role against radiation in human cancer cells.

Apoptosis pathways

Apoptosis is defined as programmed cell death, pathways leading to apoptosis are found to be defective in cancer.

Mitochondrial pathway: A number of stimuli, such as DNA damage can induce transcription of p53, which modulates transcription of a number of members of the Bcl-2 family BH3-only proteins. These proteins translocate to the mitochondria where they promote the release of cytochrome c and/or inhibit anti-apoptotic bcl-2/bcl-xl. Cytochrome c then binds to Apaf-1, which further complexes with caspase-9 to form the apoptosome and promotes cleavage of effector caspases.

Death receptor pathway: Upon Ligand binding, Fas/Trail receptors aggregate and recruit via FADD adaptor molecules a number of caspase-8 molecules, this close proximity allows for mutual cleavage resulting in their activation. Caspase-8 can cleave effector caspases such as caspase-3 resulting in a caspase cleavage cascade. Decoy receptors DcR1/DcR2 compete with trail receptors for Trail binding. DcR1/DcR2 do not propagate the caspase-signalling cascade due to non-functional or absent intracellular domains. Crosstalk between pathways occurs at the caspase level. Caspase-8 can cleave Bid, whereby its C-terminal translocates to the mitochondria and promotes cell death via activation of Bax and Bak. Cleavage of effector caspases (by either pathway) results in apoptosis induction and its associated phenotype (DNA fragmentation, membrane blebbing, cell shrinkage and the formation of apoptotic bodies).

Reactive oxygen species (ROS) and mitochondria play an important role in apoptosis induction under both physiologic and pathologic conditions. Interestingly, mitochondria are both source and target of ROS. Cytochrome C release from mitochondria, that triggers caspase activation, appears to be largely mediated by direct or indirect ROS action. On the other hand, ROS have also anti-apoptotic effects.

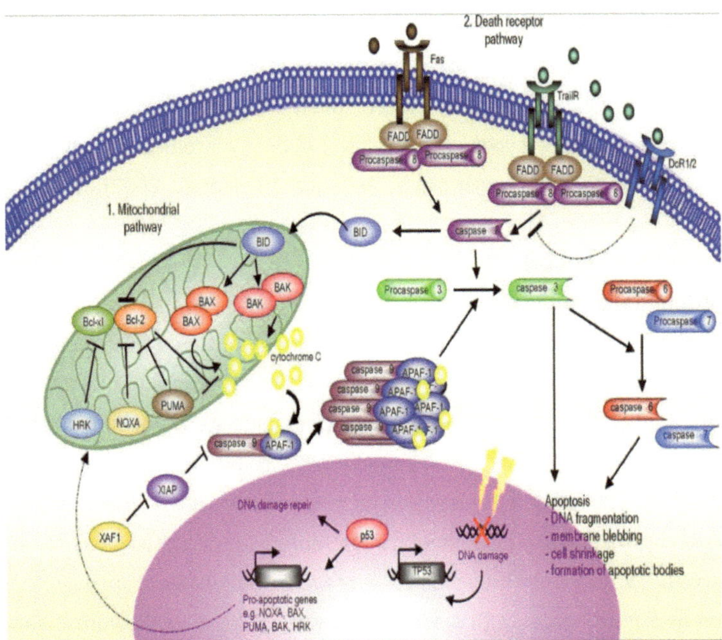

Figure 3. Apoptosis process (*Reference: Endocrine related cancer Harmones and cancer: From research to therapies*)

Cellular response to radiation

DNA repair systems and cell cycle checkpoints closely co-operate in the attempt of maintaining the genomic integrity of cells damaged by ionizing radiation. DNA double-strand breaks (DSB) are considered as the most biologically important radiation-induced damage. Their spatial distribution and association with other types of damage depend on radiation quality. It is believed these features affect damage reparability, thus explaining the higher efficiency for cellular effects of densely ionizing radiation with respect to gamma-rays. DSB repair systems identified in mammalian cells are homologous recombination (HR), single-strand annealing (SSA) and non-homologous end-joining (NHEJ). Some enzymes may participate in more than one of these repair systems. DNA damage also triggers biochemical signals activating checkpoints responsible for delay in cell cycle progression that allows more time for repair. Those at G1/S and S phases prevent replication of damaged DNA and those at G2/M phase prevent segregation of changed chromosomes. Individuals with lack or alterations of genes involved in DNA DSB repair and cell cycle checkpoints exhibit syndromes characterized by genome instability and predisposition to cancer (Belli et al., 2002).

Materials and Method

Experimental animal as mice

Animals belonging to a conventional inbred colony of Swiss mice were selected for the experiment. Male mice (8 weeks) weighing 20-25g and maintained on a standard laboratory diet and water. All experiments were conducted with strict adherence to the ethical guidelines laid down by the "Committee for the Purpose of Control and Supervision of Experiments on Animals" (CPCSEA), Government of India. The animals were reared in air conditioned room (24 ± 2^0C) with 12h dark light schedule; they were then housed in polypropylene cages (4-5 mice/ cage) with sterile paddy husk. The animals were maintained with on a diet containing 70% cracked wheat, 20% cracked Bengal gram, 5% fish meal, 4% yeast, 0.25% shark liver oil, 0.75% seas am oil containing 21%

protein,64.8% carbohydrate, 5% of fats, 5% of fibre besides 4.1% mineral & vitamins.

Irradiation of mice

The animal were divided into two groups

1. Control (unirradiated)

2. Whole body irradiated (cobalt 60) in perpsex box designed for BARC workshop.

Doses (0.1Gy, 2Gy and 10 Gy)

Time period (1h, 3h, 7h, 1day and 3days).

Isolation of splenocytes

Mice spleen, Disposable Pasteur pipettes (sterile) Polystyrene Petri dish, clean scalpel (optional), Cell strainer (70 mm Nylon),   2 ml syringe, sterile tubes (30 ml and 50 ml), serological pipettes of appropriate volumes(sterile).

Ammonium chloride lysing reagent (0.15 MNH4Cl, 1 mM KHCO3, 0.1 mM EDTA, or a commercial), phosphate buffered saline (PBS), wash buffer (0.1% sodium azide, 0.1% BSA inPBS) Bench top centrifuge (NOT refrigerated) with swing-out rotor and appropriate carriers, hemacytometer and microscope for cell counting

Procedure

1. Prepare ammonium chloride lysing reagent. Leave at room temperature.

2. Place a cell strainer in the Petri dish. Transfer the spleen and 1 ml PBS directly into the cell strainer. If desired, score the outer membrane of the spleen with a clean scalpel before mashing it, but take care to avoid cutting through the strainer mesh.

3. Remove the plunger from a 2 ml syringe and use the black rubber end to mash the spleen and release the splenocytes into the Petri dish. Use grinding circular movements to homogenize the tissue. Periodically, draw up liquid from outside the strainer with a disposable pipette, and wash out the cells from within the strainer. Continue to mash the spleen until all that remains is the white connective tissue of the outermembrane.

4. Transfer the homogenized cell suspension into a universal tube. Wash out the Petri dish a few times to maximize recovery of splenocytes. Make up to the full volume of the tube with PBS.

5. Centrifuge at 400 × g for 10 minutes at room temperature and aspirate supernatant. The resulting cell pellet should be red in color. The supernatant may appear cloudy, due to smaller particles that cannot be centrifuged at this setting.

6. Resuspend the cell pellet in 2 ml PBS per spleen. Add the appropriate amount of ammonium chloride lysing reagent (1 ml per 100 ml blood, or as defined by the manufacturer) and leave for 15minutes in the dark at room temperature. Ensure that the cells are fully resuspended before adding the lysis buffer in order to avoid excessive clumping.

7. Centrifuge at 400 × g for 5 minutes and aspirate the supernatant. Take care not to lose cells as thepellet will be loose. The cell pellet should be buff colored with minimal RBC contamination.

8. Resuspend the cells completely and wash again with a full volume of PBS. Centrifuge at 400 × gfor 5 minutes, aspirate supernatant, then resuspend cells in desired medium (e.g. wash buffer orRPMI-1640) to a final known volume.

9. Count live cells using a hemacytometer and light microscope. Resuspend the cells in a few ml of wash buffer and use the same day.

Estimation of protein by bicinchoninic acid method

BCA Protein Assay is a detergent-compatible formulation based on bicinchoninic acid (BCA) for the colorimetric detection and quantitation of total protein. This method combines the well-known reduction of Cu^{+2} to Cu^{+1} by protein in an alkaline medium (the biuret reaction) with the highly sensitive and selective colorimetric detection of the cuprous cation (Cu^{+1}) using a unique reagent containing bicinchoninic acid.

1.The purple-colored reaction product of this assay is formed by the chelation of two molecules of BCA with one cuprous ion. This water-soluble complex exhibits a strong absorbance at 562nm that is nearly linear with increasing protein concentrations over a broad working range (20-2000µg/mL). However, following incubation, the rate of continued color development is sufficiently slow to allow large numbers of samples to be assayed together.The macromolecular structure of protein, the number of peptide bonds and the presence of four particular amino acids (cysteine, cystine, tryptophan and tyrosine) are reported to be responsible for color formation with BCA.

2 Studies with di-, tri- and tetrapeptides suggest that the extent of color formation caused by more than one functional groups. Accordingly, protein concentrations generally are determined and reported with reference to standards of a common protein such as bovine serum albumin (BSA). A series of dilutions of known concentration are prepared from the protein and assayed alongside the unknown(s) before the concentration of each unknown is determined based on the standard curve. Bovine gamma globulin (BGG) used as protein standard. Test Tube Procedure requires a larger volume (0.1mL) of protein sample; however, because it uses a sample to working reagent ratio of 1:20 (v/v), the effect of interfering substances is minimized.

Preparation of Standards and Working Reagent (required for both assay procedures)

A. Preparation of Diluted Albumin (BSA) Standards

Use Table 1 as a guide to prepare a set of protein standards. Dilute the contents of one Albumin Standard (BSA) ampule into several clean vials, preferably using the same diluent as the sample(s). Each 1mL ampule of 2mg/mL Albumin Standard is sufficient to prepare a set of diluted standards for either working range suggested in Table 1. There will be sufficient volume for three replications of each diluted standard.

Table 1. Preparation of Diluted Albumin (BSA) Standards

Dilution Scheme for Standard Test Tube Protocol and Microplate Procedure (Working Range = 20-2,000µg/mL)

Vial	Volume of Diluent (µL)	Volume and Source of BSA (µL)	Final BSA Concentration (µg/mL)
A	0	300 of Stock	2000
B	125	375 of Stock	1500
C	325	325 of Stock	1000
D	175	175 of vial B dilution	750
E	325	325 of vial C dilution	500
F	325	325 of vial E dilution	250
G	325	325 of vial F dilution	125
H	400	100 of vial G dilution	25
I	400	0	0 = Blank

Dilution Scheme for Enhanced Test Tube Protocol (Working Range = 5–250µg/mL)

Vial	Volume of Diluent (µL)	Volume and Source of BSA (µL)	Final BSA Concentration (µg/mL)
A	700	100 of Stock	250
B	400	400 of vial A dilution	125
C	450	300 of vial B dilution	50
D	400	400 of vial C dilution	25
E	400	100 of vial D dilution	5
F	400	0	0 = Blank

B. Preparation of the BCA Working Reagent (WR)

1. Use the following formula to determine the total volume of WR required:

(# standards + # unknowns) × (# replicates) × (volume of WR per sample) = total volume WR required Example: for the standard test-tube procedure with 3 unknowns and 2 replicates of each sample: (9 standards + 3 unknowns) × (2 replicates) × (2mL) = 48mL WR required

2. Prepare WR by mixing 50 parts of BCA Reagent A with 1 part of BCA Reagent B (50:1, Reagent A:B). For the above example, combine 50mL of Reagent A with 1mL of Reagent B.

Note: When Reagent B is first added to Reagent A, turbidity is observed that quickly disappears upon mixing to yield a clear, green WR. Prepare sufficient volume of WR based on the number of samples to be assayed. The WR is stable for several days when stored in a closed container at room temperature (RT).

Procedure (Sample to WR ratio = 1:20)

1. Pipette 0.1mL of each standard and unknown sample replicate into an appropriately labelled test tube.

2. Add 2.0mL of the WR to each tube and mix well.

3. Cover and incubate tubes at selected temperature and time:

• Standard Protocol: 37°C for 30 minutes (working range = 20-2000µg/mL)

• RT Protocol: RT for 2 hours (working range = 20-2000μg/mL)

• Enhanced Protocol: 60°C for 30 minutes (working range = 5-250μg/mL)

Notes:

• Increasing the incubation time or temperature increases the net 562nm absorbance for each test and decreases both the minimum detection level of the reagent and the working range of the protocol.

• Use a water bath to heat tubes for either Standard (37°C incubation) or Enhanced (60°C incubation) Protocol. Using a forced-air incubator can introduce significant error in color development because of uneven heat transfer.

4. Cool all tubes to RT.

5. With the spectrophotometer set to 562nm, zero the instrument on a cuvette filled only with water. Subsequently, measure the absorbance of all the samples within 10 minutes.

Note: Because the BCA assay does not reach a true end point, color development will continue even after cooling to RT. However, because the rate of color development is low at RT, no significant error will be introduced if the 562nm absorbance measurements of all tubes are made within 10 minutes of each other.

6. Subtract the average 562nm absorbance measurement of the Blank standard replicates from the 562nm absorbance measurement of all other individual standard and unknown sample replicates.

7. Prepare a standard curve by plotting the average Blank-corrected 562nm measurement for each BSA standard vs. its concentration in μg/mL. Use the standard curve to determine the protein concentration of each unknown sample.

Western blotting

Lysis buffers

These buffers may be stored at 4°C for several weeks or for up to a year aliquoted and stored at -20°C.

Nonidet-P40 (NP40) buffer

150 mM NaCl

1.0% NP-40 (possible to substitute with 0.1% Triton X-100)

50 mM Tris-HCl pH 8.0

Protease Inhibitors

RIPA buffer (Radio Immuno Precipitation Assay buffer)

150 mM NaCl

1.0% NP-40 or 0.1% Triton X-100

0.5% sodium deoxycholate

0.1% SDS (sodium dodecyl sulphate)

50 mM Tris-HCl pH 8.0

Protease Inhibitors

Tris-HCl buffer

20 mM Tris-HCl pH 7.5

Protease Inhibitors

Running, Transfer, and Blocking buffers

loading buffer

4% SDS

10% 2-mercaptoethanol

20% glycerol

0.004% bromophenol blue

0.125 M Tris-HCl

Check the pH and adjust pH to 6.8.

Running buffer (Tris-Glycine/SDS)

25 mM Tris base

190 mM glycine

0.1% SDS

Check the pH, which should be about pH 8.3. Adjust if necessary.

Transfer buffer (Wet)

25 mM Tris base

190 mM glycine

20% methanol

Check the pH, which should be about pH 8.3. Adjust if necessary.

For proteins larger than 80 kDa, we recommend that SDS is included at a final concentration of 0.1%.

Transfer buffer (Semi-dry)

48 mM Tris

39 mM glycine

20% methanol

0.04% SDS

Blocking buffer:

5% milk or BSA (bovine serum albumin)

Add to TBST buffer. Mix well and filter. Failure to filter can lead to "spotting" where tiny dark grains will contaminate the blot during color development.

Procedure:

1. Sample lysis

Preparation of lysate from cell culture

Place the cell culture dish in ice and wash the cells with ice-cold PBS.

Aspirate the PBS, then add ice-cold lysis buffer (1 ml per 107 cells/100 mm dish/150 cm2   flask; 0.5ml per 5x106 cells / 60 mm dish / 75 cm2 flask).

Scrape adherent cells off the dish using a cold plastic cell scraper, then gently transfer the cell suspension into a pre-cooled microcentrifuge tube. Maintain constant agitation for 30 minutes at 4°C. Spin at 16,000 x g for 20 minutes in a 4°C pre-cooled centrifuge. Gently remove the tube from the centrifuge and place on ice. Transfer the supernatant to a fresh tube kept on ice, and discard the pellet.

Preparation of lysate from tissues

Dissect the tissue of interest with clean tools, on ice preferably, and as quickly as possible to prevent degradation by proteases.Place the tissue in round-bottom microcentrifuge tubes or Eppendorf tubes and immerse in liquid nitrogen to "snap freeze". Store samples at -80°C for later use or keep on ice for immediate homogenization. For a ~5 mg piece of tissue, add ~300 µl lysis buffer rapidly to the tube, homogenize with an electric homogenizer, rinse the blade twice with another 2 x 300 µl lysis buffer, then maintain constant agitation (for example, on an orbital shaker) for 2 hours at 4°C. Centrifuge for 20 minutes at 16,000 x g at 4°C in a microcentrifuge for 20 minutes. Gently remove the tubes from the centrifuge and place on ice. Transfer the supernatant to a fresh tube kept on ice. Discard the pellet.

2. Sample preparation

Remove a small volume (50 µl) of lysate to perform a protein assay. Determine the protein concentration for each cell lysate.To the remaining volume of cell lysate, add an equal volume of 2X Laemmli Sample Buffer.

To reduce and denature: Boil each cell lysate in sample buffer at 100°C for 5 minutes and aliquot. Store lysates at -20°C. Note: aliquot cell lysates (50- 100 µl) to avoid repeat freeze/thaw cycles. Defrost tubes containing cell lysate at 37°C. Centrifuge at 16,000 x g in a microcentrifuge for 5 minutes.

3. Loading and running the gel

Load equal amounts of protein into the wells of the SDS-PAGE gel, along with molecular weight markers. Load 20- 30 µg of total protein from cell lysate or tissue homogenate, or 10 -100 ng of purified protein. Run the gel for 1 to 2 hours at 100 V.

This time and voltage may require some optimization.

Gel percentage will depend on the size of the protein:

4 - 40 kDa	20%
12 - 45 kDa	15%
10 - 70 kDa	12.5%
15 - 100 kDa	10%
25 - 200 kDa	8%

4. Transferring the protein from the gel to the membrane

Figure 5: Prepare the transfer stack as above given order

Western blot transfer

The membrane can be either nitrocellulose or PVDF; each has its advantages. "Activate" PVDF with methanol for one minute and rinse with transfer buffer before preparing the stack. The time and voltage may require some optimization. Transfer to the membrane can be checked using Ponceau Red staining before the blocking step. The membrane is ready for antibody staining.

5. Antibody staining

Block the membrane for 1 hour at room temperature or overnight at 4°C using 5% blocking solution. Incubate membrane with appropriate dilutions of primary antibody in 5% or 2% blocking solution overnight at 4°C or for 2 hours at room temperature. Wash the membrane in three washes of TBST, 5 minutes each.
Incubate the membrane with the recommended dilution of labelled secondary antibody in 5% blocking buffer in TBST at room temperature for 1 hour. Wash the membrane in three washes of TBST, 5 minutes each, then rinse in TBS. Remove excess reagent and cover the membrane in transparent plastic wrap. Acquire image using darkroom development techniques for chemiluminesence, or normal image scanning methods for colorimetric detection.

The proteins under study were

P44, p38, PKC (α,β,δ)

Dilution used for primary antibody (1:1000) for each protein rabbit IgG cell signalling

Dilution used for secondary antibody (2:1000) for each protein rabbit IgG cell signalling prepared in blocking solution.

Develop the blot using chemiluminsence method.

Results and Discussion

Expression of signalling molecules at 10cGy

Expression of p44/42: 10cGy which is considerably low dose to be used in radiotherapy, it has been observed that at 1h there was significant decrease in p44/42,but thereafter 3h there was increase in levels. As p44/42 are involved in cytoprotective response of the cell. In spleen which is a radiosensitive organ there should be decrease in the cytoprotective response which finally leads to apoptosis. As mentioned earlier 10cGy is too low dose used for radiotherapy, so difference in trend of expression is insignificant (fig 5).

Expression of PKCα,β: The trend observed in expression of these protein were increase at 1h, and decrease at 3h, 7h, 24h and 72h similar to control. The cells are known to undergo apoptosis after 6-7h of irradiation.

Expression of PKCδ: The expression of protein was significant at 1h but less than control, with marginal increase at 3h and decrease at 7h. The expression at 24h was observed negligible and abruptly increased at 72h. As PKCδ is substrate to caspases hence there is decrease in the levels at 24h and increase at 72h.

Expression of signalling molecules at 2Gy

The expression of p44/42 increases at 1h, marginal decrease at 3h and drastic increase at 7hfollowing negligible expression at 72h. Expression of PKCα,β decreased with time 3h and 7h and expression of PKCδ increased with time. As 2Gy is relevant dose used for radiotherapy, the expression of cytoprotective protein were low in radiosensitive organ (fig 6).

Expression of signalling molecules at 10Gy

There was decrease in expression of cytoprotective proteins p44/42 and PKCα,β and increase in concentration of PKCδ. As 10Gy is high dose cell undergo death(fig 7) .

Expression of signalling molecules p38

Increase in level of p38 can be observed which is stress induced signalling molecule indulge cell into apoptosis (fig 8).

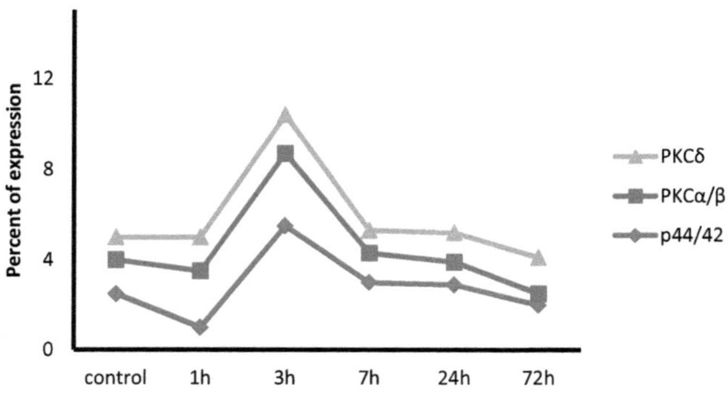

Figure 5. Expression of signalling molecule at 10cGY of radiation.

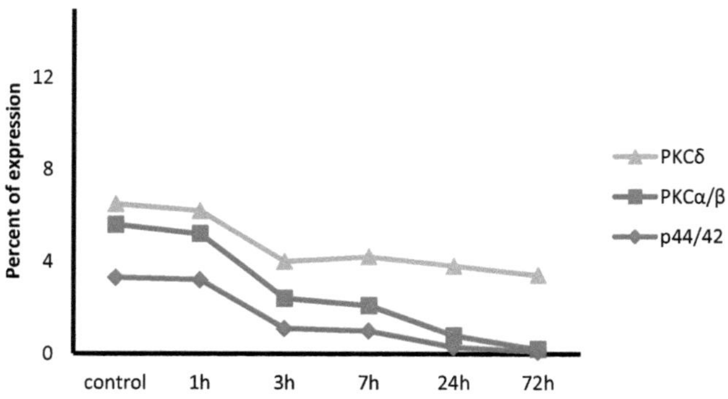

Figure 6. Expression of signalling molecule at 2GY of radiation.

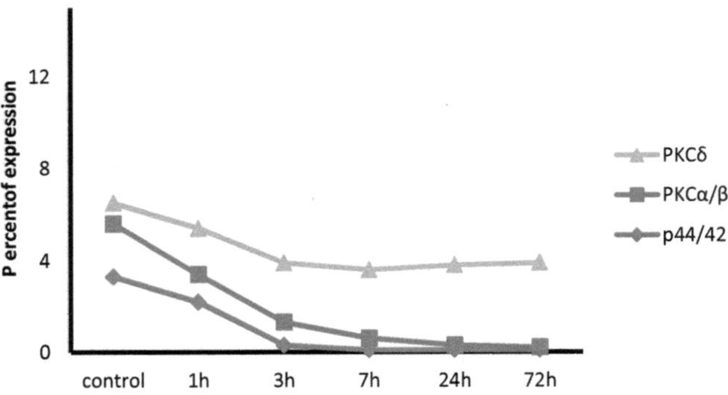

Figure 7. Expression of signalling molecule at 10GY of radiation.

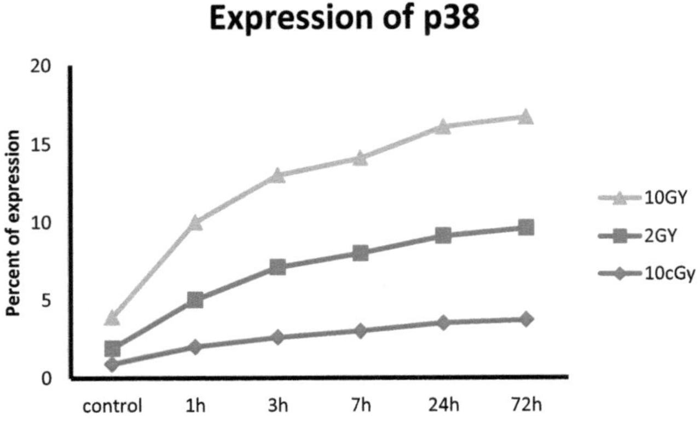

Figure 8. Expression of p38 signalling molecule at 10cGY,2GY and 10GY of radiation.

References

Altman A, Villalba M Protein kinase C-theta (PKCtheta): it's all about location, location, location. *Immunological reviews* 192:53-63(2003).

Baan B,van Dam H,van der Zon GC,Maassen JA,Ouwens DM.The role of JNK,p38 and ERK MAP-kinases in insulin-induced Thr69 and Thr71-phosphorylation of transcription factor ATF2.*Mol Endocrinol.* (2006)

Beibei Cai, Sandra H. Chang, Esther B. E. Becker, Azad Bonni and Zhengui Xia p38 MAP Kinase Mediates Apoptosis through Phosphorylation of BimEL at Ser-65*.*THE Journal of biological chemistry* 281, NO. 35, pp. 25215–25222,(2006).

Belli M, Sapora O, Tabocchini MA. Molecular targets in cellular response to ionizing radiation and implications in space radiation protection.*J Radiat Res.*43 Suppl:S13-9 (2002).

Biological Effects of Radiation. Reactor concept manual. USNRC Technical Training Center page: 9-17.

Dalby, K.N. et al. *J Biol Chem* 273, 1496-505 (1998).

Dorn GW 2nd, Force T Protein kinase cascades in the regulation of cardiac hypertrophy. *The Journal of clinical investigation* 115(3):527-37(2005).

Fang X, Yu S, Tanyi JL, Lu Y, Woodgett JR, Mills GB Convergence of multiple signalling cascades at glycogen synthase kinase 3: Edg receptor-mediated phosphorylation and inactivation by lysophosphatidic acid through a protein kinase C-dependent intracellular pathway. *Molecular and cellular biology* 22(7):2099-110(2002).

Hamilton M, Liao J, Cathcart MK, Wolfman A Constitutive association of c-N-Ras with c-Raf-1 and protein kinase C epsilon in latent signalling modules. *The Journal of biological chemistry*;276(31):29079-90(2001).

Heasley LE,Han SY.JNK Regulation of Oncogenesis. *Mol Cells.* 30;21(2):167-73. (2006)

Himes SR,Sester DP,Ravasi T,Cronau SL,Sasmono T,Hume DA. The JNK are important for development and survival of macrophages. *J Immunol.*15;176(4):2219-28 (2006).

Hug H. and Sarre T.F.. Protein kinase C isoenzymes: divergence in signal transduction. *Biochem. J.* 291, 329-343 (1993).

Kortenjann, M. et al. *Mol Cell Biol* 14, 4815-24 (1994).

Kraus S,Benard O,Naor Z,Seger R.c-Src is activated by the epidermal growth factor receptor in a pathway that mediates JNK and ERK activation by gonadotropin-releasing hormone in COS7 cells.*J Biol Chem*.29;278(35):32618-30. Epub (2003).

Li J, O'Connor KL, Hellmich MR, Greeley GH Jr, Townsend CM Jr, Evers BM. The role of protein kinase D in neurotensin secretion mediated by protein kinase C-alpha/-delta and Rho/Rho kinase. *The Journal of biological chemistry* 2;279(27):28466-74 (2004).

Liu W.S. and Heckman C.A. The sevenfold way of PKC regulation. *Cell. Signal.* 10, 529-542 (1998).

Matsukawa J,Matsuzawa A,Takeda K,Ichijo H.The ASK1-MAP kinase cascades in mammalian stress response. *J Biochem* (Tokyo);136(3):261-5 (2004).

Meloche, S. and Pouysségur, J. *Oncogene* 26, 3227-39. (2007)

Meng Zhao • Li Xia • Guo-Qiang Chen Protein Kinase Cd in Apoptosis: A Brief Overview Arch. *Immunol. Ther. Exp.* 60:361–372(2012)

Moulin N,Widmann C. Islet-brain (IB)/JNK-interacting proteins (JIPs): future targets for the treatment of neurodegenerative diseases. *Curr Neurovasc Res*.1(2):111-27 (2004).

Murphy, L.O. and Blenis, J. *Trends Biochem Sci* 31, 268-75. (2006)

Newton A.C. Regulation of protein kinase C. *Curr. Opin. Cell Biol.* 9, 161-167. (1997).

Owens, D.M. and Keyse, S.M. *Oncogene* 26, 3203-13. (2007)

Parra M, Kasler H, McKinsey TA, Olson EN, Verdin E Protein kinase D1 phosphorylates HDAC7 and induces its nuclear export after T-cell receptor activation. *The Journal of biological chemistry* 8;280(14):13762-70 (2005).

Radiation effects research foundation. A co-operative Japan-US research organisation. http://www.rerf.jp/radefx/basickno_e/radcell.htm

Sanz L, Sanchez P, Lallena MJ, Diaz-Meco MT, Moscat J The interaction of p62 with RIP links the atypical PKCs to NF-kappaB activation. *The EMBO journal* 1;18(11):3044-53 (1999).

Sprowles A,Robinson D,Wu YM,Kung HJ,Wisdom R.c-Jun controls the efficiency of MAP kinase signalling by transcriptional repression of MAP kinase phosphatases. *Exp Cell Res*.15;308(2):459-68 (2005)

Storz P, Doppler H, Toker AProtein kinase Cdelta selectively regulates protein kinase D-dependent activation of NF-kappaB in oxidative stress signalling. *Molecular and cellular biology* 24(7):2614-26 (2004).

Storz P, Toker A Protein kinase D mediates a stress-induced NF-kappaB activation and survival pathway. *The EMBO journal* 2;22(1):109-20 (2003).

Wang YM, Seibenhener ML, Vandenplas ML, Wooten MW Atypical PKC zeta is activated by ceramide, resulting in coactivation of NF-kappaB/JNK kinase and cell survival. Journal of neuroscience research 1999 Feb 1;55(3):293-302

Wooten MW, Vandenplas ML, Seibenhener ML, Geetha T, Diaz-Meco MT Nerve growth factor stimulates multisite tyrosine phosphorylation and activation of the atypical protein kinase C's via a src kinase pathway. *Molecular and cellular biology* 21(24):8414-27 (2001).

Yamauchi J,Miyamoto Y,Kokubu H,Nishii H,Okamoto M,Sugawara Y,Hirasawa A,Tsujimoto G,Itoh H.Endothelin suppresses cell migration via the JNK signalling pathway in a manner dependent upon Src kinase,Rac1,and Cdc42.*FEBS Lett.* 527(1-3):284-8 (2002).

Yang L,Mao L,Chen H,Catavsan M,Kozinn J,Arora A,Liu X,Wang JQ.A signalling mechanism from G alpha q-protein-coupled metabotropic glutamate receptors to gene expression: role of the c-Jun N-terminal kinase pathway. *J Neurosci.* 18;26(3):971-80 (2006).

Yang Q,Kim YS,Lin Y,Lewis J,Neckers L,Liu ZG.Tumour necrosis factor receptor 1 mediates endoplasmic reticulum stress-induced activation of the MAP kinase JNK.*EMBO Rep* (2006).

Zemlickova E, Johannes FJ, Aitken A, Dubois T Association of CPI-17 with protein kinase C and casein kinase I. *Biochemical and biophysical research communications* 26;316(1):39-47(2004).

Zhou JY,Liu Y,Wu GS.The role of mitogen-activated protein kinase phosphatase-1 in oxidative damage-induced cell death. *Cancer Res*1;66(9):4888-94 (2006).

Zhou JY,Liu Y,Wu GS.The role of mitogen-activated protein kinase phosphatase-1 in oxidative damage-induced cell death. *Cancer Res.* 66(9):4888-94 (2006).